THE WOMAN AND THE CAR

Photo by Foulsham & Banfield, Ltd.

THE WOMAN AND THE CAR

A CHATTY LITTLE HANDBOOK FOR ALL WOMEN WHO MOTOR OR WHO WANT TO MOTOR

BY DOROTHY LEVITT

EDITED WITH INTRODUCTORY ARTICLES BY C. BYNG-HALL

ILLUSTRATED BY PHOTOGRAPHS SPECIALLY TAKEN

Published in Great Britain in 2014 by Old House books & maps
c/o Osprey Publishing, PO Box 883, Oxford OX2 9PH, UK.
c/o Osprey Publishing, PO Box 3985, New York, NY 10185-3985, USA.
Website: www.oldhousebooks.co.uk

A CIP catalogue record for this book is available from the British Library.

ISBN-13: 978 1 90840 287 5
Originally published in 1909 by John Lane, the Bodley Head.
Printed in China through Worldprint Ltd.

Extra illustrations are acknowledged as follows:
Mary Evans Picture Library, pages 147, 149 and 150;
Topfoto, page 152.

14 15 16 17 18 10 9 8 7 6 5 4 3 2 1

INTRODUCTORY

In presenting this book to the public the publisher is acting largely on the request of some hundreds of ladies, some already motorists, others would-be motorists. Miss Dorothy Levitt, last year, wrote a short series of articles for the *Daily Graphic* on the subject of Motoring for Women. These articles attracted a great deal of attention and Miss Levitt was inundated with letters from all parts of the United Kingdom and also from abroad, asking her for further information on various points and also begging her to publish the articles and additional information in volume form.

Miss Levitt was also asked to contribute articles on the same lines to many magazines and weekly publications and further received requests from a number of distinguished women to give them personal instruction in the art of driving and managing the mechanism of their cars.

As the simplest way out of answering all these requests Miss Levitt has revised and enlarged her former articles and has added new chapters and a great deal of matter which she believes every woman motorist or beginner will find of use.

There has been no attempt to make this volume a formal text-book on motoring for women but rather a chatty little handbook, containing simple and understandable instructions and hints for all women motorists, whether beginners or experts.

The facts contained in the various chapters are not those gathered from any standard manual of motoring but are from Miss Levitt's own practical experience of six years' daily driving, in all sorts of cars, in all sorts of weather and under all sorts of conditions—pleasure trips, long-distance tours at home and abroad and in competitions.

There may be points here and there which she has overlooked. Miss Levitt, however, will answer such questions or furnish such further information as readers may properly desire, either through the medium of his Majesty's mails or, perhaps, in a later edition of this volume.

The photographs, with which the several chapters are illustrated, were specially taken for the work by Mr. Horace W. Nicholls.

London, February 1909.

CONTENTS

ILLUSTRATIONS

Illustrations

Illustrations

DOROTHY LEVITT: A PERSONAL SKETCH

It is not considered difficult for mere man to write about a pretty, young woman. Yet in the case of Dorothy Levitt it is difficult. There are so many things in her delightful private life which would have a vivid interest for the public. But I am forbidden to tread too deeply in that direction.

Dorothy Levitt is the premier woman motorist and botorist of the world. And she is ready to prove and uphold her title at any time.

In the United Kingdom, in France and in Germany, she has achieved distinctions, won success and carried off trophies such as no woman and few men can claim.

Five years ago Miss Levitt won the Championship of the Seas in the great motor-boat race at Trouville, France, defeating all comers.

Three years ago at Brighton she won a race and created a world's record for women of 79¾ miles per hour. The following year she broke her own record and created a new world's record for women of 91 miles an hour.

Looking at Miss Levitt one can hardly imagine that she could drive a car at such terrific speed. The public, in its mind's eye, no doubt figures this motor champion as a big, strapping Amazon. Dorothy Levitt is exactly, or almost so, the direct opposite of such a picture. She is the most girlish of womanly women. Slight in stature, shy and shrinking, almost timid in her everyday life, it is seeming a marvel that she can really be the woman who has done all that the records show.

And the way in which she came to be a motorist—it is a story in itself. She was from childhood a good cyclist, a good driver of horses, a rider to hounds and an excellent shot with rifle or gun. Fishing was her favourite pastime. She was quick of eye and sure of hand and nerves troubled her not at all.

A friend, owning a motor-car, paid a visit

to the family in the West Country. In a very few days Dorothy Levitt had become well acquainted with the intricacies of that motor. She handled the wheel as well as the owner or his chauffeur. She attended, as a spectator, a county competition, driving the car with such skill that the attention was attracted of the manager of a big motor firm. He secured an introduction and asked her to drive one of his cars in a competition. She agreed and thus became the first Englishwoman to drive a motor-car in a public competition.

Her first prize was won a month later, and since then she has steadily mounted the tree of her chosen profession. Yet she has remained an amateur, accepting no money prizes, only medals and cups and such like trophies.

In hill climbs, endurance and speed trials she is alike invincible. At the first aerial hare-and-hounds race of balloons this year she was selected as the umpire. The most careful, as well as intrepid and fast-driving motorist, was wanted. Miss Levitt unerringly followed the hare from London to near Arundel, Sussex, and was on the spot when the first balloon among the hounds descended near the hare.

Miss Levitt has been offered many enticing professional engagements on the Continent and in the United States but prefers to remain at home and an amateur.

In appearance Dorothy Levitt looks partly French, partly Irish, with a *soupçon* of American. Yet she is wholly English. Of medium height, her figure is slim and very graceful. She has a very girlish but expressive face, large eyes that are brown and grey and green in varying lights, brown hair that curls, a straight nose that has the bare inclination of a saucy upward tip and a mouth which is too large. It is a charming, winning face.

The one fault of Dorothy Levitt is her modesty, almost amounting to bashfulness. One cannot get her to tell much of her many exciting adventures, particularly those of which she is the heroine. She is immensely popular, has been toasted by Royalty at German motor banquets, elected honorary member of many of the first automobile clubs in this country and on the Continent, and has a host of friends, some in the sacred circles of society, others distinguished men and women of the more Bohemian circles of art and literature, music and the drama. She is an inveterate first-nighter, wears simple but ravishing clothes and, to those who do not know her, passes as a bright butterfly of fashion.

In a flat in a quiet but fashionable neighbourhood in the West-end of London, Miss Levitt lives the life of a bachelor girl. There she has a housekeeper and maid and a tiny Pomeranian, one "Dodo," to keep her company. The flat contains, as its feature rooms, a Louis XIV. drawing-room and a Flemish dining-room, the latter the scene of many little luncheon parties for which Miss Levitt is also famous.

Hers is a busy life, involving many thousands of miles of travel in the year. She is to be seen at Ascot, Goodwood, Cowes, at Henley, at Ranelagh. To-day she may be in London. Next week you may hear of her as in France or Germany taking part in a motor competition; the week following she may be in Scotland or of a house-party in the Shires or botoring in the blue waters off the Riviera coast.

It is little wonder that her ambition is to leave the gay whirl and to settle down quietly in the country, with her motor, her dogs and a fishing-rod and a gun.

Of her public records I can do no better than quote extracts from her diary, for Miss Levitt, unlike the generality of women, is most careful in keeping a very businesslike diary. Here are the extracts:

April 1903.—First Englishwoman to take part in public motor-car competition. Did not win. Will do better next time.

May 13, 1903.—Glasgow to London Non-stop Run. Drove 16 horse-power Gladiator. Gained 994 marks out of possible 1000. Marks deducted for tyre troubles.

August 1903.—Won Gaston Menier Cup at Trouville, France. Value of cup, 350 guineas.

August 8, 1903.—Drove motor-boat *Napier* at Cowes. Won the race. Afterwards commanded to go over to Royal yacht by the King as his Majesty wanted to see me put boat through its paces. King thinks such boats may be useful for despatch work.

September 1903.—One thousand miles Reliability Trials. Sixteen horse-power Gladiator. Won. Did fastest time in class.

October 2, 1903.—Southport Speed Trials. Drove 16 horse-power Gladiator. Won silver cup for speed.

Won Championship of the Seas, Trouville. Napier motor-boat. Boat afterwards bought by French Government for £1000.

September 1904.—Light Car Trials. Successful. Drove small De Dion, 8 horse-power. Entirely alone. No mechanic attended to car. Did everything myself. Had non-stop for five days but small difficulties on sixth and last day.

October 1904.—Southport Speed Trials. Drove 50 horse-power Napier. Won two medals.

February 1905.—Did Liverpool and back to London in two days, averaging a level 20 miles per hour throughout for the entire 411 miles. Unaccompanied by mechanic. Eight horse-power De Dion.

May 1905.—Won Non-stop Certificate at Scottish Trials. Ran over very rough and hilly roads in the Highlands. Eight horse-power De Dion.

July 1905.—Won Brighton Sweepstakes on 80 horse-power Napier, at rate of $79\frac{3}{4}$ miles per hour, constituting the woman's world record. Beat a great many professional drivers. Drove at rate of $77\frac{3}{4}$ miles in *Daily Mail* Cup.

June 1906.—Shelsley Walsh Hill Climb. Was only sixth at finish. Fifty horse-power Napier. Mine was only car competing which was not fitted with non-skids. Car nearly went over embankment owing to this and greasy state of roads.

June 1906.—South Harting Hill Climb. Won medal on 50 horse-power Napier. Also presented with silver casket for winning private match on same hill.

July 1906.—Aston Hill Climb (Tring). Third on 50 horse-power Napier.

October 1906.—Broke my own record and created new world's record for women at Blackpool. Ninety horse-power six-cylinder

Napier. Racing car. Drove at rate of 91 miles per hour. Had near escape as front part of bonnet worked loose and, had I not pulled up in time, might have blown back and beheaded me. Was presented with a cup by the Blackpool Automobile Club and also a cup by S. F. Edge, Limited.

May 1907.—Bexhill, Second Prize, Appearance Competition. Eight horse-power De Dion.

June 1907.—Germany. Won Gold Medal Herkomer Trophy Race (1818 kilometres). Fourth out of 172 competitors. In hill climb, fifth, and tenth in Forstenrieder Park Speed Trial out of 172 competitors. Was first of all women in all competitions. Sixty horse-power six-cylinder Napier. There were 42 cars with much larger engines than I had.

October 1907.—France, Gaillon Hill Climb. Forty horse-power six-cylinder Napier. Won in my class by 20 seconds. Gradient of hill 1 in 10 average.

June 1908.—Prinz Heinrich Trophy, Germany. Made absolute non-stop run on

45 horse-power Napier. Won large silver placque.

July 1908.—Aston Hill Climb, Aston Clinton. Made second fastest time of over 50 competitors on 60 horse-power Napier.

August 1908.—France. Trouville, La Côte du Calvaire.

THE WOMAN AND THE CAR

Photo. H. W. Nicholls.

"DRIVE YOUR OWN CAR"

THE WOMAN AND THE CAR

CHAPTER I

THE CAR—ITS COST, UP-KEEP AND ACCESSORIES

Motoring as a Pastime for Women—Patience of more Value than Nerve—Selection of a Car—Single-cylinder the best for Women who are going to drive themselves and attend to the Mechanism—Cost of a Small Car—Necessary Accessories and their Cost—Expense of Up-keep—The necessary Licences and the Cost

PATIENCE, the capacity for taking pains, is of more value than the most ponderous nerve. You may be afraid, as I am, of driving in a hansom through the crowded streets of town—you may be afraid of a mouse, or so nervous that you are startled at the slightest of sudden sounds—yet you can be a skilful motorist, and enjoy to the full the delights of this greatest of out-door pastimes, if you possess patience—the capacity for taking pains.

Motoring is a pastime for women: young,

middle-aged, and—if there are any—old. There may be pleasure in being whirled around the country by your friends and relatives, or in a car driven by your chauffeur; but the real, the intense pleasure, the actual realisation of the pastime comes only when you drive your own car.

I have hunted—and was one with those who declare that the most glorious of all out-door life is in the saddle, on a fast, clean-jumping hunter; but when, by accident, I took up motoring I found the exhilaration, the delights of the gallop doubled. It fascinated me, and it will fascinate any woman who tries it.

I am writing this little book not so much for those women who have already taken up motoring, but for those who would like to, but either dare not because of nervousness, or who imagine it is too difficult to understand the many necessary details.

In the following chapters I will endeavour to explain everything in the simplest possible manner, without lapsing into confusing technicalities.

The first thing to discuss is the car. There

are scores of makes, good, bad and indifferent. I have tried many different makes and have come to the conclusion that the De Dion is an ideal single-cylinder car for a woman to drive. It combines simplicity with reliability—two very important items to the automobiliste.

For your own driving, if you are going to attend to the mechanism yourself, you should purchase a single-cylinder car—more cylinders mean more work, and also more expense as regards tyres, petrol, oil, &c. The single-cylinder car is the most economical to run. Being constructed in a much lighter manner the weight on the tyres is less, consequently the tyre bill is smaller, a matter of great importance in the upkeep of a car.

The horse-power of a single-cylinder car is usually 8 h.p. or less. As regards carriage work, of course the purchaser can suit herself, but the "Victoria" type of body has the most graceful lines. Colour, also, is a matter of one's own selection. Dark blue, brown, green, red or cream, they all look well, and can be picked out with lines to match the upholstery, or further embellished with a top panel of basket-work, as is the car in the photograph.

Such a car as I have described will cost, new, from £230. This price, however, is for the car itself, upholstered and complete as to seats and side lamps. It is the accessories that bring up the cost. It adds greatly to one's comfort to have a hood, made of either black leather or khaki-coloured canvas, with nickel or brass mountings to match the finish of your car. Such a hood will cost, in leather, about £20, and in canvas £18. I am quoting for the best quality in every instance, for with motoring it is quality that counts in the long run. A folding glass screen, with nickel or brass fittings, framed in stained wood, will cost £10. The front lamps will cost about £6 per pair, and the rear lamp £1 to £1 5*s*. A waterproof rug can be bought for £1 to £2.

The car will, of course, seat two, but it is often advisable to have a third seat. This should be constructed so as to fold down when not in use, and would cost £15. You can have a stationary seat fitted for £10, but these do not look so nice (though quite as comfortable)

Photo. H. W. Nicholls

IT IS ACCESSORIES THAT BRING UP THE COST—YOU MUST HAVE A HOOD

as those that fold down. In addition to these things it is necessary to carry a tyre repair outfit, which will cost about £1, also the following tools and spares :

Ammeter, jack, pliers, spanners, carburetter jet key, large and small screw-drivers, hammer, oil-can, grease injector, tyre-pump, sparking-plug, inlet and exhaust valves, trembler blade and screw, some washers, split pins, file, very fine file for platinum points, emery-powder, insulated tape, and some waste or swabs.

In buying your car you will probably find that the last car you were on is " the best." It is liable to become somewhat confusing if you go for many trial runs ; but one thing to bear in mind is that the car that will do five miles an hour faster than the one you previously tried is not necessarily the best car—it may be faster while it is running, but it may not run for long—therefore take my advice and pin your faith on the car with the reputation for reliability, the one that will not entail a big expenditure every few months for repairs. Nearly all of us, nowadays, have some motoring friends, who have probably had experience with different makes of cars : their experience should benefit you in your choice. There are some very inexpensive cars on the market, but inexpensive only as to initial outlay—they are likely to prove themselves sorry bargains before many months have passed. One of the chief joys of motoring is to feel that you can rely upon your car.

In regard to housing the car, if you are not fortunate enough to possess a stable or garage of your own, one of the following courses is open to you: Hire a stable, or garage, and a man to attend to the cleaning of the car; place it at one of the many garages, public and semi-private, now in existence—or you can stable it at the nearest mews and arrange with the ostler to do the washing. I cannot give the exact cost of the first and last of these, as they would naturally vary, but if placed at a regular garage the cost would be from 8*s.* 6*d.* to 12*s.* 6*d.* per week.

Another outlay is to be found in "tips." The men at a garage are always hungry for "tips," and your car will be polished with greater zest if the "tips" are frequent or generous. The advertisement of the "no tip" garage is a fallacy. The proprietor may consider this principle the right one, but if you act according to his ideas your car will probably suffer.

Petrol varies slightly in price, but is usually from 1*s.* to 1*s.* 4*d.* per gallon. As, with the car illustrated, you are able to run 28 to 32 miles on one gallon, you will see that petrol is not a great item.

The next duty that devolves on you after becoming the owner of a car is to procure your licences. There are two—one a licence for the car (the same as a carriage licence), the cost of which is governed according to the weight of the car, probably near two guineas; the other a driving licence, costing 5*s.* Both of these are to be obtained from the London County Council offices in Spring Gardens, London, S.W., or in the country at the various County Council headquarters—though the carriage licence can be obtained, after due application, at almost any post office. Your driving licence is an official printed paper with your name and address written in.

To obtain a number for the car it is necessary to apply to Spring Gardens, or any of the Registration and Licensing authorities in the United Kingdom. If one writes to a County Council, the letter should be addressed: "Clerk to the County Council of ——, County Council Offices, ——," and if to a county borough, to "The Town Clerk, Town Hall, ——." The registration fee is 20*s*. You will have to fill up a form and will then have a number registered.

This number you must have painted on two tin plates, white on a black ground. The figures must be 3½ in. deep. These number-plates must be affixed to the back and front of your car. The back one must be so placed that the light from the back lamp is thrown on it and thus the number distinctly seen at night. It is also necessary for part of the rear lamp to show a red light.

You should never go in your car without this licence—your driving licence, for you must produce it when asked by the proper authorities or pay the penalty of £5. But more of this later.

CHAPTER II

THE ALL-IMPORTANT QUESTION OF DRESS

The All-important Question of Dress—Masks and Goggles are usually unnecessary—"Nothing like Leather" is a False Cry—The best Head-gear—A Neck-muffler is of the greatest Importance—Beware of Rings and "fluffy" Things—The Question of the Overall—What the Secret Drawer should contain—Hints about all Garments—Suggestion anent carrying a Revolver

An all-important question is dress. Automobilists are nowadays more careful in the choice of their attire, but there are still a goodly number who seem to imagine it is impossible to look anything but hideous when in an automobile. On a closed-in car, limousine or landaulette, any kind of attire is permissible as the conditions are precisely the same as being driven in a carriage, but with an open car neatness and comfort are essential. When racing, or when in countries where speed is not looked upon with such horror as in England

—on the long, straight seemingly never-ending, military roads of France, one can travel at a speed that makes goggles or masks a necessity, but for motoring under ordinary conditions there is no reason why one should wear them. It might be borne in mind that I am writing this book for the woman who is desirous of being her own driver and owning her own car; yet perhaps my advice will be applicable to the whole sex. I average about 400 miles per week—in all conditions of weather—all sorts of cars and all sorts of places, and therefore speak from experience—in many instances dearly bought.

Now, as to ordinary garments, dress for the season of the year exactly as you would if you were not going motoring. I would advise shoes rather than boots as they give greater freedom to the ankles and do not tend to impede the circulation, as a fairly tightly laced or buttoned boot would do, but this is a matter of individual taste. In winter time it is advisable to wear high gaiters, have them specially made, almost up to the knee.

As regards a frock—the plain "tailor-made"

Photo. H. W. Nicholls.

ONE OF THE MOST IMPORTANT ARTICLES OF WEAR IS A SCARF OR MUFFLER FOR THE NECK

with a shirt blouse of linen, silk or " Viyella " is without doubt the most comfortable—and the wearer has the advantage, at the end of a days' run, of appearing trim and neat. Under no circumstances wear lace or " fluffy " adjuncts to your toilette—if you do, you will regret them before you have driven half a dozen miles.

Regarding coats—there is nothing like a thick frieze, homespun, or tweed, lined with " Jaeger " or fur. The former has the advantage of being lighter in weight than the latter and is just as warm and much less expensive. In England in winter one can wear a coat of this description right up to the beginning of summer. For summer itself, the ideal coat is of thin cream serge. It retains its freshness and does not crease like alpaca, linen or silk. The serge looks, and feels, smart all the summer —the silk or alpaca, after its first hard day, begins to look creased and shabby.

Do not heed the cry " nothing like leather." Leather coats do not wear out gracefully. At first they may be delightful, but when they have been caught in two or three showers they begin to have a hard, stiff feeling which is far from comfortable. I have, however, seen very pretty costumes, coats and skirts, made of thin glove kid, or *suède*, but these are luxuries, as they cost from twenty-five to thirty guineas each.

As to head-gear, there is no question: the round cap or close-fitting turban of fur are the most comfortable and suitable, though with the glass screen up it is possible to wear an ordinary hat, with a veil round it. However, if you go in for caps, see that they fit well—there is nothing more uncomfortable than the cap that does not fit. It is a good plan to have caps made to match your costumes. When fixing the cap, pin it securely, and over it put a *crêpe-de-chine* veil, of length a-plenty. These can be obtained from most of the leading drapers, and it is quite a simple matter to make them yourself with a length of *crêpe* or washing silk. Before tying the veil, twist the ends. This prevents the knot working loose and is very necessary, as the veil, in addition to protecting the hair, helps to keep the hat securely in place.

Photo. H. W. Nicholls

REMEMBER TO TWIST THE VEIL BEFORE TYING—THIS PREVENTS THE KNOT WORKING LOOSE

One of the most important articles of wear is a scarf, or muffler, for the neck—and the manner of wearing it is also important. Fold it, then wind round the throat, beginning at the front, bringing the ends round from the back, and fold over in front. See that the throat is covered closely, and not too loosely. Wearing this properly will save you all manner of colds, sore throats and kindred sufferings.

Regarding gloves—never wear woollen gloves, as wool slips on the smooth surface of the steering-wheel and prevents one getting a firm grip. Gloves made of good, soft kid, fur-lined, without a fastening, and made with just a thumb, are the ideal gloves for winter driving.

It is not advisable to wear rings. If you do not want to leave them at home, or in a hotel, but want to wear them when you are indoors, during your ride or tour take them off while on the car and stow them away. Rings, when you are driving yourself, hurt terribly, and also the stones are loosened. Bracelets and bangles are irritating unless secured by a sleeve or glove from working up and down.

Indispensable to the motoriste who is going

to drive her own car is the overall. This should be made of butcher-blue or brown linen, to fasten at the back—the same shape as an artist's overall. It should have long sleeves. You can always slip off your coat and put on the overall in a moment—and it is necessary if you have anything to do in the car. Remember it is better to get grease-spots on your washable overall than on your coat or other clothes.

While there are several little repairs that it would be impossible to remedy if wearing gloves, the majority of work on a car (filling tanks, &c. &c.) can be done just as well if one's hands are protected by a pair of wash-leather gloves. You will find room for these gloves in the little drawer under the seat of the car.

This little drawer is the secret of the dainty motoriste. What you put in it depends upon your tastes, but the following articles are what I advise you to have in its recesses. A pair of clean gloves, an extra handkerchief, clean veil, powder-puff (unless you despise them), hair-pins and ordinary pins, a hand mirror—

Photo. H. W. Nicholls.

"THE USEFUL OVERALL"

and some chocolates are very soothing, sometimes!

It is also advisable to carry a tablet of "Antioyl" soap. If it has been necessary to use bare hands for a repair you will nearly always find some grease on your hands, and this it is impossible to remove with ordinary soap. Of course it is possible to remove it with a little petrol, but I have found that petrol roughens the skin and that the "Antioyl" soap is much better.

The mirror should be fairly large to be really useful, and it is better to have one with a handle to it. Just before starting take the glass out of the little drawer and put it into the little flap pocket of the car. You will find it useful to have it handy—not for strictly personal use, but to occasionally hold up to see what is behind you. Sometimes you will wonder if you heard a car behind you—and while the necessity or inclination to look round is rare, you can, with the mirror, see in a flash what is in the rear without losing your forward way, and without releasing your right-hand grip of the steering-wheel.

If you are going to drive alone in the highways and byways it might be advisable to carry a small revolver. I have an automatic "Colt," and find it very easy to handle as there is practically no recoil—a great consideration to a woman. While I have never had occasion to use it on the road (though, I may add, I practise continually at a range to keep my eye and hand "in") it is nevertheless a comfort to know that should the occasion arise I have the means of defending myself.

If you are driving alone a dog is great company. The majority of dogs like motors and soon get into the habit of curling up on the seat by your side, under your coat.

Photo. H. W. Nicholls.

THIS LITTLE DRAWER IS THE GREAT SECRET

CHAPTER III

THE MECHANISM OF THE CAR

In which a Practical Introduction to the Car is given with Simple Explanations of the Details of the Machinery—The Importance of Lubrication—The Testing of the Brakes—The Six Levers and their Various Functions—The Electric Battery

"Be sure you are right, then go ahead." This good old motto is just the thing to remember when one is going in for motoring. Remember, I am discussing the woman who drives her own car, and does all those things that ordinarily a chauffeur would have to do.

I am constantly asked by some astonished people, "Do you really understand all the horrid machinery of a motor, and could you mend it if it broke down?" but it really is not a very difficult matter. The details of the engine may sound complicated and may look "horrid," but an engine is easily mastered. A few hours of proper diligence, provided you are determined to learn, and you know all that you have to know. Again, I must remind you that I am discussing the single-cylinder car, which is by far the simplest for a woman to drive and attend to alone.

I have made it a rule never to allow any one to drive my own little car—and this is a rule that every one will find useful. All cars have their individual idiosyncrasies, and if you alone drive, you get to understand every sound; but if you allow any one to drive you are ignorant of what strain the car has been put to. As a matter of fact, a strange hand on the wheel and levers seems to put the car out of tune.

Before starting out for a ride your first duty is to see that the petrol-tank is full. It is unpleasant to be stranded on the road, miles from anywhere, minus petrol. The petrol-tank is, in many instances, under the seat. Lift the cushions, unscrew the cap and peep in. If it is dark it will be necessary to hold a piece of stick in to see how much petrol there is, but when there is occasion to do this, be very careful that there is no dirt on the stick,

Photo. H. W. Nicholls

UNSCREW THE CAP AND PEEP IN

or a choked petrol-pipe or carburetter will result. The slightest atom of dirt in the petrol will cause trouble. If you are going for a very long run it is a wise plan to take an extra can with you. It is, perhaps, unnecessary for me to warn you not to take a light near the petrol-tank while it is being filled up. Many cars have been wrecked through carelessness in this direction. Remember it is not actually the *petrol* that catches alight, but the vapour that arises from it. If your petrol-tank runs dry there is no danger—the car will simply come to a standstill.

The chamber in which the petrol and air mix and vapourise is called the carburetter, from which the vapour is carried to the cylinder head by means of a pipe, and is there exploded by the tiny electric spark from the sparking-plug, the explosion forcing down the piston and causing the crank-shaft to revolve.

Having examined your petrol-supply, being sure to replace the screw quite tightly, your next duty is to see that the water-tank is full. This tank is situated behind the engine, close to the dash-board. Unscrew the top and if you cannot see without uncomfortably bending over take a twig or stick and poke it in, and the wet portion will tell you how much water there is ; though on some cars there is, affixed to the screw tops of both the petrol- and water-tanks, a metal rod which shows the amount of petrol or water in the tank. When you are refilling the water-tank you can tell by the " overflow " when the tank is full—there is a special outlet, so as to prevent the surplus from flowing over the top of the engine. Be sure to screw the top of the tank on again.

The next for examination is the oil-tank. This supplies the oil to lubricate the engine and gears. There are several different types of lubricators, force or drip feed, according to the type of car. It is necessary to lift the bonnet to refill the oil-tank. The " bonnet " is the metal covering to the engine. When the bonnet is lifted, metal supports will be found each side to hold it up. It is important that you have plenty of oil, for it is the lubrication that keeps your machinery in working trim. Without oil your engine and

Photo. H. W. Nicholls.

TEST THE QUANTITY OF OIL IN THE TANK BY INSERTING A PIECE OF STICK

gears would overheat and probably "seize." The lubricating oil is forced, by a small hand pump, to all the different parts of the engine and gears requiring it.

With the car illustrated, it is necessary to pump a charge of oil into the engine about every twenty miles. This is an easy matter and it is not necessary to stop the car to do it. Before starting out each day you should allow the " used " oil to run out of the base chamber. This is done by lifting a small rod you will find on the left-hand side of the commutator. If you pull this up it releases the oil, which you will see running out. When it is all out do not forget to press the rod into place again, as failure to do this would mean serious trouble, as the fresh oil, when pumped into the engine, would simply run right through on to the ground. After this stale oil has been released, two charges should be pumped into the engine before starting. This is done by turning the pointer on the pump handle to " Reservoir," then pull up slowly, turn the pointer to " Moteur " and press gently down. To lubricate the gear, fill from " Reservoir " as before, turn the pointer to "Vitesse" and press down in the same manner.

Apart from filling the various grease-caps occasionally, on the steering, &c., and greasing the wheel bearings, this is all that is necessary in the way of lubrication. The wheels, however, only require greasing about every 400 miles.

Your next duty would be to test the brake. Get into the habit of doing this every time you go out. It is no trouble to run the car a few yards to ascertain whether the brakes grip or not. If all motorists, no matter how expert, were to spend a few moments in taking this precaution, there would undoubtedly be fewer accidents. We often read that "the steering-gear went wrong," but I am right in saying that, in many cases, the accidents are caused by the failure of the brakes when applied suddenly on an emergency. The brakes may be in a satisfactory condition when you lock up the car after a day's run, but when the car is stationary there is the slight possibility of a little oil dripping on to them during the night, rendering them practically useless. If there

Photo. H. W. Nicholls.

PULL UP THIS SMALL ROD TO ENABLE THE "USED" OIL TO RUN OUT OF THE BASE-CHAMBER

is oil on the brakes, however, it can easily be burnt off by running the car a little way and applying the brakes several times—the friction will gradually burn it off. The brakes sometimes slacken and need adjustment—an easy operation. Types of brakes may vary slightly with different types of cars On my single-cylinder car the brake is very simple. To adjust, you will find a small handle underneath the frame towards the back of the car, on the off-side. Half a turn one way or the other if the brake is too slack or too fierce usually suffices. It can be adjusted one way or the other in about a second or two.

We now come to the various levers and their different functions. There are six levers, the change-speed, or gear-lever, on the left-hand side of the steering-column, under the steering-wheel; the ignition-lever and the air-lever, both to be found under the steering-wheel on the right-hand side of the steering-column; the hand-throttle, underneath the steering-wheel; in centre of column, on a small ratchet, the hand-brake lever and the first-speed lever. In this chapter I shall only deal with the first four.

It is the gear-lever that sets the gear in motion—practically puts the "muscles" of your car into play. It is imperative that it be in neutral (or central) position when starting the engine, or when applying either the hand- or foot-brake. This gear-lever, when in neutral position, disconnects the engine and gear, thereby allowing the road wheels to be brought to a standstill. When first learning to drive there is a tendency to suddenly apply the brake without bringing the gear-lever into neutral position, consequently a great strain is put upon the entire mechanism, as rival forces are brought into play, viz., while the engine is pulling, and through the driving-shaft turning the gears and propelling the wheels, the action of the brake is to make the back wheels stationary. As an example, the shock on the mechanism in the event of this happening can be compared to taking a watch and banging it on the ground.

With regard to the other levers, always

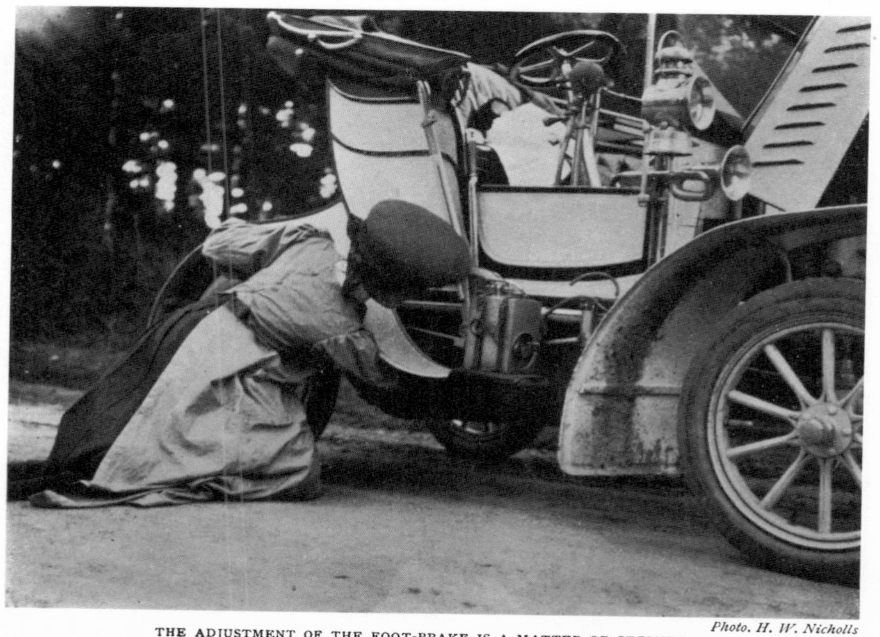

Photo. H. W. Nicholls

THE ADJUSTMENT OF THE FOOT-BRAKE IS A MATTER OF SECONDS

be quite sure that the ignition (top) lever is retarded, and that the air-lever is in correct position. It is impossible to exactly give the correct position for these as it is a matter of adjustment and liable to slightly vary with different cars. However, when once you have found the correct position it is impossible to get it wrong.

The next thing to do is to switch on the electric current. The car carries a battery on the dash-board, where is also the coil. As it is possible to run 2000 or 3000 miles on one battery, the replacing is not an expensive item. The cost of a new battery is 15*s.* 9*d.* The switch is on the coil and the current is set in motion by moving the switch from " A " to " M." Having done all these things you are ready to start up the engine, and after that, drive.

It has taken a long time to explain the preliminary things that you should do before starting off on a run, and the non-motoriste, and even perhaps the intending motoriste, will say, " If I have to do all those troublesome things it will take up all my time, so I think

I had better have a chauffeur"; but let me assure you that while it has taken some little time to explain these things in the plainest possible language, it will take you but a few minutes to carry them out.

CHAPTER IV

HOW TO DRIVE

Starting the Engine—How to hold the Steering-wheel—Various Speeds and Gears—How to start the Car—The Art of Throttling—The Use of Foot-pedals—Changing Speed—How to climb Hills—Running Downhill and on the Level—How to use Brakes—Skidding—Driving Backwards

> "*Fire in the heart of me, moving and chattering,*
> *Youth in each part of me, slender and strong,*
> *Light and tremendous I bear you along.*"

THESE lines, I feel sure, appeal to every motoriste as they exactly describe the little car in motion.

Starting one's engine is not the nicest thing about motoring when a woman is acting as her own driver and mechanician, but with the little cars no strength is required; it is only the big cars that need a swelling of muscle. There is a great knack in starting an engine, but this, once overcome, ceases to be hard work.

In front of the car you will notice a handle. Push it inwards until you feel it fit into a notch, then pull it up sharply, releasing your hold of the handle the moment you feel you have pulled it over the resisting (compression) point. Unless starting a car fitted with magneto ignition, on no account press down the handle—always pull it upwards, smartly and sharply. If it is pressed down the possibility of a backfire is greater—and a broken arm may result. This, however, is not a common occurrence, and is one that is brought about entirely through carelessness on the part of the would-be driver.

If the car has not been used for some hours it will sometimes be necessary to turn the starting handle two or three times—speaking from my own experience, three times is the maximum; it will usually start on the second turn. Of course in the winter it will take two or three turns, as the petrol freezes and takes longer to vapourise.

The moment the engine is running you can get in the car and start driving. Hold the steering-wheel with both hands in the

Photo. H. W. Nicholls.

IN FRONT OF YOUR CAR YOU WILL NOTICE A HANDLE

manner and position in which it is most comfortable to yourself. Keep a firm hold of it all the time and do not get into the habit of driving with one hand on the wheel : use both except when it is necessary to use either for changing speed, &c. Advance the ignition-lever forward and give more air by pulling back the " air " lever. When you have tried the car once or twice you will easily be able to gauge the distance these various levers should be moved. Remember the faster the engine runs the more air she will take, though when climbing a hill it is necessary to drive on a richer mixture (less air, with the regular supply of petrol) and cut off the air inlet almost entirely.

You will soon understand the different sounds of the engine—and their prevention, or cure.

When the car is stationary (and the engine running) always endeavour to run it as quietly as possible. It is sometimes annoying to people to have the noise of an automobile outside their door and no good can accrue by allowing the engine to run on unchecked. Underneath the steering-wheel, on the steering-column you will notice the throttle-lever, mentioned in the preceding chapter. Its function is to regulate the mixture in the carburetter. When you wish the engine to run very quietly and slowly, you move this little lever from right to left. When starting to drive again do not omit releasing the throttle, otherwise the engine may stop owing to insufficiency of gas.

Your next move is to take off the side brake. You will find this lever on the right-hand side of the car. Now you are absolutely ready to start. Always remember that a car should receive careful treatment, so therefore do not attempt to move anything by jerking it roughly—take everything very quietly.

You will notice two pedals on the left and right respectively of the steering-column, on the floor. The left-hand pedal acts as a throttle (shutting off gas) in exactly the same manner as the hand-throttle explained above, that is, when it is pressed down half-way—and the throttling process acts according to the amount of pressure put upon the pedal.

Photo. H. W. Nicholls

RELEASE THE FOOT FROM THE RIGHT PEDAL AND THROTTLE SLIGHTLY WITH THE LEFT FOOT ON THE LEFT PEDAL

If it is pressed down still further it acts as a brake. This foot-throttle and foot-brake are more frequently used than the hand-throttle and hand-brake.

The small pedal on the right-hand side of the steering-column is brought into service when first (or lowest) speed is required. Thus, to start the car you press down this pedal as far as it will go, at the same time pressing lightly on the throttle (left-hand pedal), and take hold of the change-speed (or gear) lever and push it as far forward as it will go, at the same time releasing the left-hand pedal and keeping down the one on the right. You are now travelling on first speed. You will soon tell when this speed has reached its maximum power as the engine will be running very fast. It will then be time to change to second speed. This is done by releasing the right-hand pedal, throttle slightly with the left foot and bring the gear-lever towards you as far as it will go, at the same time slowly lifting your foot off the throttle. You will now find the car moving much faster and will be able to tell, as in the former instance, when the maximum speed is obtained. When you hear the engine beginning to " race " (run very fast), slightly throttle again and push the gear-lever into third speed—away from you, the same position as for " first," though, of course, in this instance it is unnecessary to touch the right-hand pedal, which, you must remember, is only used for lowest speed. You are now on third (or top) gear and will find that unless the country is very hilly you will usually be able to keep the car running on this gear for a long time, varying the speeds from 10 to 28 miles an hour by deftly manipulating the ignition, air and throttle.

Changing speed on the car illustrated is an easy task, and that is why I have taken the De Dion car as an object-lesson. With the majority of cars there is a deal of " arm work " necessary, and in many cases not a little strain, this strain very often counteracting the benefits otherwise derived from the driving.

In changing speed always remember to throttle slightly, whether in changing from first or second to top, or in slackening speed, from top to second or first.

Never change from first to top speed, or from top to first without using the intermediate speed. The first speed on these little cars is from zero to 9 miles an hour, the second from 9 to 18, and the top from 18 to 28. I should advise you to get thoroughly used to the steering while on second speed, and at first drive very slowly. Do not expect to be able to control the car in a few minutes. Take your time and get in sympathy with your motor as you would the horses you drive or ride. Gain confidence slowly. Once you have confidence in yourself the battle is nearly won. Bear in mind that when riding or driving a horse it is only partly under your control. As it has a brain and will of its own it can bolt if it wishes to—but with a motor-car you rely upon yourself alone—you are master (or should I say mistress ?) of the situation.

When it is necessary to drive backwards, push forward the small lever you will find at the side, by the hand-brake, push the change-speed lever forward (same position as for top speed) and steer exactly the reverse way to what you would do if you were going forward.

This will no doubt prove awkward at first, and will necessitate a little practice, but when once mastered you will find it equally easy to steer the car either in a backward or forward direction. When running on any speed, if you allow the car to gain impetus and then put the change-speed lever in neutral position, the car will travel on in the same manner as a free-wheel bicycle, the action of the gear being neutral temporarily disconnecting the engine and gear.

If you see an obstacle in the road do not go up to it at full speed and suddenly put on the brake. The sudden application of the brake will hasten the end of the life of your tyres—and if you are not driving on non-skid tyres a bad accident may be the result if the road is at all wet. Of course occasion may sometimes arise whereby it is absolutely necessary to apply the brakes very suddenly—for instance, if a car makes its appearance from a side turning that perhaps you had not noticed, or an individual or dog attempts to cross the road a few yards in front of your car—but in cases of this kind you must rely on your own judg-

Photo. H. W. Nicholls.

YOUR NEXT MOVE IS TO TAKE OFF THE SIDE BRAKE

ment. You will soon understand how far the car will run with a given impetus and learn to use the brakes gently and sparingly.

Never get into the habit of using only one brake. It is more convenient to use the foot-brake, but if you do not sometimes use the hand-brake, when an emergency arises you will find yourself looking for the hand-brake instead of your hand instinctively going out to it in the fraction of a second.

In travelling uphill run as far as possible at top speed—meanwhile listening to the throb of the engine. When the top speed drops to about the maximum speed of the second, it is time to change back to second, but do not allow the engine to run too slowly before changing, otherwise it will not "pick up" and it will be a great strain on it. If the hill is a very steep one it may be necessary to change to first. There is always a correct moment for changing speed, both on the level and on hills, but this is a thing that can only come with practice.

If you have a long decline to negotiate, turn the switch off so that the engine is not running, allowing the car to roll down on its own impetus, controlling with the brakes, but remember, on the car slowing down, after passing the foot of the hill, to put the top gear in and switch on again, otherwise the engine may stop.

If you are driving in a very hilly or mountainous country you must give the engine a charge of oil more frequently than every twenty miles on account of its having to be on low gear, when the engine runs much faster and is liable to get over-heated—and if it does become over-heated you will soon notice a nasty "knock"—regarding which I will explain in a later chapter, on "Troubles."

CHAPTER V

TROUBLES—HOW TO AVOID AND TO MEND THEM

The Battle of Motor Woe—Various Troubles which may happen at any Time—How to diagnose each Trouble and how to repair it—Prevention better than Cure—In spare Time, Practice in Repairing is Valuable—Simple Instructions from the Tightening of a Nut to the putting on of a Tyre

YOUR troubles with a car may be *nil*—they may be a-plenty. You may be at fault, and again, the trouble may simply be one of ordinary misfortune or due to the idiosyncrasies of your car : but to whatever it is due, learn quickly to mend matters and laugh at them rather than weep. I well remember the first time I started out alone without a chauffeur. Somehow or other the car stopped (cars in those days were not so reliable as they are to-day—and the one I had lent me had done good service). For several hours I could not make out what was the matter, wept bitter tears and was so down-hearted that it took me a day to get over it: but most of the little troubles that arise are easily overcome on a modern car, when once one understands how to locate them. Train your ear to distinguish the slightest sound foreign to the consistent running of the engine. A single misfire means that there is some little thing needing attention; but always remember to switch off the electric current before touching anything—if you do not you will get a shock.

There is only one trouble regarding which you are really justified in feeling angry—that is a punctured or burst tyre. It *is* possible for a woman to repair a tyre, but I am sure I am correct in saying that not one woman in a thousand would want to ruin her hands in this way. Nowadays there is a repairer in nearly every village, and the best plan is to drive very slowly on the rim to this nearest repair-shop. With a small, light, single-cylinder car it will do no harm, but if you drive too fast, or far, the inner tube may be damaged. However, it is advisable to carry a "Stepney" wheel. These wheels are really indispensable and should have the place of honour on a woman's

Photo. H. W. Nicholls.

IT IS A SIMPLE MATTER TO REMOVE A FAULTY SPARKING PLUG

car. This "Stepney" wheel is an ordinary wheel, fitted with flanges to fix on to the existing wheel, and carries a tyre already pumped up—and can be affixed to your car in less than ten minutes. No strength is required to put it in place.

In regard to tyres—there are several good makes, such as Dunlop, Continental, Michelin, and several others. The stock car you buy from the maker will probably be fitted with one of these makes, but will have plain or corrugated treads. As there is such a great danger through skidding it is advisable to pay a little extra and have non-skid tyres fitted. In addition to preventing the car turning round on a greasy road, being steel-studded they will obviate puncture to a great extent I use them winter and summer, for although the country roads in summer are dry and the danger of sideslip very small, when you reach a town you will nearly always find the tram-lines (if there are any) have been watered, and it is really in towns where the non-skid tyres are a necessity.

There are a dozen little things that may

occur to you at any time, and which you can easily remedy yourself, but before starting on a lonely ride it would be well for you to practise the remedying of the troubles, in fact, give yourself lessons in them. As I have already stated, you may go almost a year without troubles of any kind; still, you should know all there is to know about them.

Sooted plug.—This is caused by the engine getting too much oil. If the plug is "sooted" it will appear to be very black and oily-looking. It is a matter of a minute to loosen the sparking-plug with a spanner, and replace it with a new one; but if you are not carrying a spare plug, and are not in a hurry for a few minutes, the dirt is easily washed off with a little petrol.

Faulty plug.—If the engine misfires it may either be owing to the above-mentioned trouble, or may be caused by the porcelain becoming cracked or loose in its seating. As this cannot be repaired a new plug is necessary.

Gap on sparking-plug badly adjusted.—If the engine is not pulling well it may be owing to the fact that the two tiny points across which the spark leaps are either too wide apart or are set too closely together. If you get a good spark with the points in a certain position it is a good plan to insert the blade of a small pen-knife before finally replacing the plug, making a mark on the knife where it was inserted. On a future occasion this will be found useful, as if the knife is again inserted between the points of a new plug and the points either opened or closed, to meet the marked portion of the blade, the correct adjustment will at once be arrived at.

To test battery.—It is necessary to do this periodically, as if the battery is not showing a sufficient number of ampères the car will run badly. The test is made with an ammeter (provision for which has been made in the list of spares), the positive points making contact with those corresponding on the battery, the ampèrage being registered on a small dial. It should always show four or over. When it gets below this point it is advisable to carry a spare battery, as a stop on the road through a battery giving out is annoying, and a battery cannot be procured at every repairer's.

Empty petrol-tank.—If the car comes to a standstill after a few misfires, it may be caused through the petrol-supply being exhausted on account of the tank having sprung a leak or the petrol merely being used up. The former is a very unusual occurrence, rarely occurring on a car with the petrol-tank under the seat. Where the tank is placed at the back of the frame of the car it is often punctured by a sharp stone striking it. Always, if going any distance, carry a spare can of petrol and a funnel, and before filling the tank be quite sure that there is no dust in the funnel or on the top of the can, as the smallest quantity of dirt will choke your petrol or carburetter, and this takes a long time to remedy.

Choked petrol-pipe or carburetter.—You may possibly believe at first that this trouble is caused by the petrol running out, so look in the tank, and if it contains petrol you may be sure that the flow is checked in some way. "Agitate" the float of the carburetter and it will most likely be found that there is no petrol coming through, therefore there must be a stoppage between the tank and the car-

Photo. H. W. Nicholls.

IT IS A SIMPLE MATTER TO ADJUST THE TREMBLER OR SCREW

buretter. At the bottom of the carburetter there is a little joint, connecting the carburetter with a small tube through which the petrol is carried. See first that the petrol-supply is turned off, then loosen the joint and take out the pipe, then turn on the petrol and see if it comes through as far as that. If it does, the trouble must be in the carburetter jet. It is a simple task to remove this with the carburetter jet spanner, and an ordinary hair-pin would then remove every obstacle.

Water in carburetter.—This may sometimes cause the engine to run badly, and is not an unusual occurrence. On the top of the carburetter being removed the water will easily be noticed. It is heavier than petrol and the little drops of water in the petrol look like drops of lead. In order to guard against suspected water you can filter the petrol through a handkerchief laid in the funnel.

Sticking valves.—With the inlet valve this trouble may be caused by oil and dirt. If this is the case it can be taken out and washed in petrol and replaced. The car may not run as it should do on account of the inlet and exhaust valves being pitted and needing grinding in. To grind in use a little emery-powder and oil, and plenty of " elbow grease " until you succeed in eliminating all the little pits and making the surface quite smooth. (Be careful to clean off all traces of the emery after grinding.)

Platinum point and screw adjustment.—The platinum point on the trembler blade or screw occasionally gets worn uneven. When contact is made, if the adjustment is not correct, the point will get worn down on one side and the engine will probably develop a " knock " and you will not be getting the maximum horse-power out of it. It is a simple matter to smooth the points, but a delicate touch is necessary. The filing must be done with your finest file, and a very few touches are necessary. Should you not happen to have a very fine file with you, the points can be smoothed with emery-paper, but care should be exercised in its use.

Again let me warn you to have the electric current switched off before making any of these repairs or adjustments, and particularly before removing the trembler blade, screw or plug. When the trembler blade and screw are replaced they will require readjustment. When you think they are set at the proper point, switch on, and give the starting-handle a few turns as if to start the engine, at the same time hold down the inlet valve, by pressing down the small button on top of cylinder. While turning, watch the spark between the two points. If it is feeble the adjustment is not correct, the screw is either tight or too loose. After a few attempts you will be able to adjust very quickly and almost unerringly.

See that the points are kept free from oil.

Loose terminals.—The car may run badly on account of a loose terminal of the wires on the sparking-plug or battery. This is simply a matter of thoroughly tightening up the various little nuts with a spanner. It is a good plan to just look over these occasionally, for when the terminals once begin to work loose they will gradually get worse and you will probably have a stop on the road.

Irregular petrol-supply.—If the carburetter

is getting too much or too little petrol, it should be remedied by readjusting the needle valve in the carburetter. When once it is properly set it will sometimes go for years without readjustment.

Punctured float.—I mention this trouble as I have personally experienced it, but it is very rare and may not occur in the whole life of a car. The float is a hollow cube, for the purpose of regulating the height of petrol in the carburetter. The float, if punctured, allows the petrol to enter, causing it to remain at the bottom of the carburetter, gradually taking in more petrol. The effect of this is that the float does not perform its office, as it becomes weighted with the petrol, and this holds open the petrol inlet, which overflows out of the carburetter.

I believe I have now enumerated all the troubles that you are likely to meet with, and which you can look after yourself, but because the list is long do not think for one moment that every time you go out in your little car you are going to experience them. You may drive for weeks, months, almost years, without

Photo. H. W. Nicholls.

BE SURE THAT ALL NUTS AND BOLTS ARE TIGHT—A RATTLE IS ANNOYING

a tiny bit of trouble, if you are careful. The looking after the little things saves a heap of trouble. The testing of this thing, the dusting of that, the tightening of a nut, the loosening of a screw—all these may be commonplace trivial matters, but if attended to will pay in the long run.

Prevention is better than cure, and the careful motoriste who looks after her car as she looks after herself will have little use for the hints in this chapter of mine.

Above all, whatever may arise, try to forget to weep and remember to laugh. Then you will have won half the battle of "Motor Woe."

CHAPTER VI

HINTS ON EXPENSES

Motoring need not be an Expensive Luxury—Two Hundred Pounds will go a long Way if properly spent—The Second-hand Car—Motor Clubs—The Ladies' Automobile Club—The Automobile Association—Motor Schools and Driving Lessons

BY the time you have read and thoroughly digested the preceding chapters I feel sure you will be able to take your car out for a spin without any misadventure.

There are but a few points which I want to impress upon you in this chapter. Do not let what you may think the great expense debar you from the pleasure of motoring. There is no great expense unless you want to make it so. In my first chapter I spoke of the price of cars and accessories. I gave a total of £300 as the average probable outlay. By no means do I want to revise these figures, but wish to remind you that the figures quoted

Photo. H. W. Nicholls.

"BE SURE THAT THE PETROL TANK IS FULL"

are, in every instance, for articles of the very best quality.

There are now being made several small cars by big firms, many of these cars being eminently suitable for a woman to drive. It is possible to procure a car at £120. The accessories, also, such as the hood and screen, need not be plated or expensively enamelled. Cape cart hoods which have the iron-work painted instead of plated are quite as serviceable, require less cleaning and the cost is considerably less ; and so with other things—but it is wise to always get the best. Durability and reliability is what you want, especially if you are limited as to expenditure. I particularly mention these matters because only the other day a friend spoke to me about the expenditure, and said that she could not possibly afford three hundred pounds. She proposed to buy a second-hand car for a small sum and have it repaired.

My advice to her was " Don't." And then I ran over a list of expenditure in getting a new car and everything necessary new. Taking the same quality as mentioned in my first chapter, but not as expensively or so well finished, I found my total less than £230.

One could, of course, go much below this by buying a second-hand car; but I would not advise this. If you know the people who have a second-hand car for sale, and can thus be assured that you will not in any way be tricked, then it might be worth while buying. But from the experience of people I know, I would rather warn you against the cars which are advertised "as good as new," and for sale for a few pounds. You would probably have to spend in repairs in the first year as much as a new car would cost.

So soon as you are the owner of a car, licensed and ready for the road, become a member of the Ladies' Automobile Club of Great Britain and Ireland. Its headquarters are situated at Claridge's Hotel, in Upper Brook Street. The club has a suite of rooms there. Send in your application to Miss K. d'Esterre Hughes, the secretary of the club.

By joining the club you have many advantages. For instance, there is, of course, the convenience of using the club rooms and the club garage when in town, and in getting a percentage off your hotel bills. But there is the greater advantage of getting all the necessary information you may want regarding hotels, roads, and such like when you want to go for a tour. There is, in fact, scarcely any information appertaining to motoring which you cannot get at the club. It is always good for a woman car-owner to belong to the first motor club in the kingdom.

Every big town has an automobile club affiliated with the Royal Club, with which the Ladies' Club is also affiliated, so that by membership in the Ladies' Club you have a standing at once with the other clubs throughout the country, and also abroad.

Every motoriste should become a member of the Automobile Association. It is an association formed for the purpose of placing scouts on the different main roads to warn motorists of police traps—and the expenditure of £2 2*s.* a year in this direction will perhaps be the means of saving you four or five times that amount within a few months. You will be given an "A.A." badge to fasten on the front of your car, and on seeing this the scouts will always stop you if there is any danger.

With regard to learning to drive, you must do so on quiet country roads or at one of the many motoring schools in and around the big cities; but know something of the school before you decide on it. There are many in which your money would be thrown away.

If you do not go to a school choose a road where there is little or no traffic. One is not allowed to learn in the parks. In fact there is quite a heavy fine imposed on inexperienced drivers who use the parks. Do not go into a street of heavy traffic until you have thoroughly mastered your car, and then drive first some half-dozen times with an expert friend as chauffeur and thus get used to the crushes and the twistings and turnings.

In traffic use your own judgment. Ladies are usually bad at judging distances, and it is well to keep as much toward the middle of the road as possible and not try too many "near things" until you have reached the expert class.

Do not be afraid to sound your horn, yet

THIS IS THE SWITCH *Photo H. W. Nicholls.*

do not use it more than necessary. At cross streets or roads and when approaching corners sound the horn and slacken speed by throttling.

There are numberless little things which, after you have graduated to the ranks of the experienced motoriste, you will buy, not because they are absolutely necessary, but because of their convenience. For instance, a speedometer. All the half-dozen makes are good ones. A speedometer is a very interesting accessory, for it tells you exactly the pace at which you are travelling, and in some instances has been known to influence the decision of a magistrate when deciding a charge of exceeding the speed-limit.

For winter driving they are now making a fur and leather arrangement which covers the steering-wheel; but I would suggest that, to the beginner at any rate, this is superfluous. Soft kid gloves, fur-lined, are much better.

Novelties of all sorts are always coming on the market; but the beginner had better let some one else try these first. It is an expensive thing to keep on experimenting with every new device for a car or the motoriste. Let others try them, and if they should prove of real use you will soon know.

Photo. H. W. Nicholls.

THIS LEVER IS USED FOR CHANGING GEAR

CHAPTER VII

MOTOR MANNERS

Things which Motoristes should and should not do when out in their Car—Laws governing the Highways—Pedestrians have the Right of Way—Points and Times at which to drive slowly—Corners and the Danger of cutting them—When to sound the Hooter—The " Courtesy of the Road "

It is, of course, quite unnecessary to teach the well-bred woman manners. The well-bred woman, with her innate courtesy and unselfishness, should she take up motoring, no doubt would act as all motoristes should act when at the wheel of their car or out on the road. So that while I have headed this chapter " Motor Manners " I desire merely to bring to the notice of readers, as prominently as possible, those things which they should do and those which they should not do when out in their cars. I can safely give this advice, for personal experience has been my teacher. For want of a better term I call these warnings and suggestions "Manners."

If every woman and man who drove a car followed these suggestions there would not be an outcry against the motor-car. Unfortunately the great majority of motorists have to suffer for other people's faults—the disgraceful driving of the few.

The laws now governing motoring have increased largely in the last few years and will probably continue in the making. I will not go into these various laws except to point out that because a person owns a motor-car the ownership of the roads is not necessarily included.

Pedestrians, according to the law, practically own the highways, not to the exclusion of other traffic, but judgments in recent cases declare that it lies with drivers to keep clear of pedestrians and that all persons have a right to walk on the highways at their own pace, whether paralytics or cripples. Dogs, chickens and other domestic animals at large on the highway are not pedestrians, and if one is driving at a regulation speed, or under, one is not responsible for their untimely end.

It is, therefore, especially advisable to drive slowly through all towns and villages. Drive slowly past all school-houses.

Always pass vehicles and bicycles on the proper side, and pass large vans, 'buses and electric tramcars very carefully, as some one may be crossing the road and suddenly appear from behind.

Drive slowly past any one driving or riding a restive horse and, if necessary, especially if it should be a lady or child riding or driving, stop the engine. This is an act of courtesy that will always be appreciated and may prevent a bad accident.

If the road is wet, give pedestrians and cyclists a wide berth so as not to splash them with mud.

Again, if the road is wet, you may be safe enough on account of your car being fitted with non-skid tyres, but in this respect the cyclist is perhaps not so fortunate. He may have a side-slip and fall perilously near your car wheels. For this reason, too, give cyclists plenty of room.

Do not fail to sound the hooter and slacken speed when coming to a cross-road, side-turning or bend. Many accidents may be averted by taking this precaution.

Never take a sharp corner at full speed. A walking pace would be much better.

Never pass or try to overtake a pedestrian, cyclist or vehicle at a corner.

Avoid the bad and perilous habit of trying to squeeze through doubtful openings in traffic either in town or country.

Never drive the engine downhill.

Do not leave the engine running when stopping outside a house. The noise, though it may be slight, may be annoying to the inmates or neighbours.

If you have a syren fitted to your car, do not sound it in a town or village. A syren is really only necessary for Continental driving.

Remember that mail-vans have the " right of way," and that ordinary traffic is supposed to give way to them.

A hooter is meant to give warning, not to startle people or wake up sleeping inmates in

Photo. H. W. Nicholls.

FIRST ADVANCE THE SPARK AND GIVE MORE AIR

their houses at all hours. Do not sound your horn oftener than absolutely necessary.

Remember, however, that it is necessary to sound the hooter when coming up behind and intending to pass a pedestrian or a vehicle. But do not wait until you are within a few feet of a pedestrian or cyclist who is already doing his utmost to get out of your way and then sound your hooter. If the cyclist be a novice or at all nervous such conduct might cause an accident.

Keep within the legal limit of speed all the time except on a good and clear stretch of road, where there happen to be no "blind" corners or dangerous cross-roads or traffic. Then there is no real harm done to any one in trying to see what you can get out of your car for a short spurt.

I cannot give you any special advice on the dust nuisance, but if you follow my suggestions, as already given, you will cause the public as little inconvenience from dust as is in your power.

There is a little thing I specially want to warn motoristes against, and that is taking corners on the wrong side of the road. Ordinarily you would not think of doing so. But wait until you come to a few corners which you can see well around. There is nothing in sight and so you skim the curb for the fun of it.

But do not keep on cutting corners—sooner or later it will become a habit and be done without thinking. Then comes the possibility of another car, a vehicle of sorts, a motorcycle or, worst of all, a cycle with a woman or child pedalling. You may not lose your presence of mind, but how about the cyclist? Don't cut corners on the wrong side of the road and there will be no need to worry about the answer to my query.

That one can show a great deal of courtesy to other cars and to general traffic on the road is assured, but that few people do is also a fact. Here is a case worthy of attention. Every motoriste has or will experience it. On the road in front of you is a covered car with noisy engine. It is a landaulette or limousine which rattles more or less. The noise of the engine is also magnified by being closed in. The car is taking up the best part of the road

Photo H. W. Nicholls.

THE AUTOMOBILE ASSOCIATION SCOUTS WILL, IF NECESSARY, STOP YOUR CAR ON THE ROAD AND

and though you are anxious to pass it you cannot, because of the noise, attract the attention of the chauffeur and get him to draw out enough for you to make a safe pass. It is very annoying and may go on for some time.

See to it, therefore, if you have a closed-in car, that there is a mirror attached to the dash-board so that the chauffeur can see what is behind him and instruct him also to keep a watch, from time to time, for coming-up cars so that you can extend to them the courtesy of the road.

One other matter may be included in "Motor Manners" and that is, leaving the car on the road or in the street unattended. In the first place the law says that you cannot leave your car unattended whether the engine is running or not. It is within the discretion of the police to summons you. They, however, do not interfere unless the engine is running noisily and the exhaust is smoking.

But in leaving one's car unattended on the road or street, care should be taken, as an act of courtesy to general traffic and pedestrians, that the car does not block the way. If on the curb in town, and it be possible, leave it on a side-street or, if in front of house or shop, give other people a chance to drive up to the front door. At the same time do not stand your car deliberately in front of some one else's house instead of your own or your friend's, if you are visiting.

Photo. H. W. Nicholls

THE ENGINE WILL START EASILY IF YOU FIRST FLOOD THE CARBURETTOR SLIGHTLY

CHAPTER VIII

TIPS—NECESSARY AND UNNECESSARY

Motoring now so general that an Owner of a Car is not any longer considered to be necessarily a Millionairess —Tipping should be on a sensible Basis—While the Motor-car has emphasised Tipping, nowadays the modest Shilling receives quite a Welcome—When to tip and when not to tip explained from Personal Experiences

If there is one thing more than another which the motor-car has revived and intensified it is the habit and practice of tipping. I need not give a lecture on tips. All of us agree, more or less, that the present-day tip is one of the banes of existence. But there are two sides to the question—one we as the givers of tips know a good deal about. Few know much about the other side—the side of the worker for and receiver of tips.

Tips must therefore be divided into two classes—the necessary tip and the unnecessary.

There are more of the latter than the former. Under the head of necessary tips I would place the garage tip, whether the garage be a public one or a private one at the house of a friend. There are a few other necessary tips, such as when a friend lends you a car for a drive or a tour or when your friend's chauffeur drives you to the railway station at an unusual hour or in very bad weather.

Luckily the motor-car is coming into such general use to-day that those who may possess one are not necessarily put down as millionaires. The chauffeur, attendants and servants generally are beginning to realise this and no longer expect a handful of money from every motoriste.

The amount of tips which should be given, in the numerous cases which I am going to mention, should depend on your income and ability to afford. That millionaires are not usually generous tippers is a well-known fact. Generally it is from the woman or man who is not very well off and who can ill afford it that the biggest tips come.

To those who count their half-crowns as worth a full thirty pence and value them accordingly, I would say—Do not be afraid to accept a friend's invitation to visit them with your motor-car because you cannot afford to do much tipping. Be sensible about this matter and I can assure you that your friend's chauffeur, or groom, will also be sensible and welcome the modest shilling or half-crown you give him.

Tipping at a public garage, if you keep your car there, has already been touched on in a previous chapter. If you go on a tour or a little trip, driving yourself, and put your car in a public garage or the one attached to your hotel or roadside inn, your car will not be touched unless you so order. Then for cleaning it, furnishing petrol, charging battery or anything else which may be wanted, there are regulation charges and these will be put down in your bill. The attendant at the garage may or may not be the man who did the work, but if he is it would be quite the proper thing to give him a small tip, just as you might tip the waiter or the chambermaid had they done any satisfactory work for you. But this need not be more than either waiter or chambermaid receive, and if your car has not been cleaned it is scarcely necessary to give the attendant even sixpence unless he has done some service for you.

Some hotels and wayside inns nowadays clean cars which stop with them overnight without extra charge, yet the fee they charge for the garage really covers this. In such case a shilling to the man who did the work would not be amiss. Your car may come into his hands again and he may do better work on it because of the little tip.

If stopping just for lunch or tea at an hotel or inn and, for convenience' sake, you run your car into the yard or garage, a small tip is necessary.

If you stop the night at a friend's house and your car is placed in your hostess's garage, you will find it spick and span in the morning with water in the tank and your petrol-tank also replenished. Perhaps this petrol has been supplied from the spare can you carry, or it may have come from your friend's supply.

You can quickly find out this. Naturally you will test your tanks and you can question

Photo. H. W. Nicholls.

THE LUBRICATION OF THE DE DION IS EXTREMELY SIMPLE

the attendant. Should the petrol-tank not be filled up and should you have used all yours you would naturally ask for enough to fill your wants. Pay for this, for in most garages nowadays a careful account is kept of petrol and other expenses. A five-shilling tip for the man is quite enough.

If your hostess should have a stable only and not a garage, and the man is only able to clean your car as he would a carriage and you have to do the filling of the tanks and the starting of the engine and so on, a smaller tip is all that is necessary.

In staying a week-end at a country house, if your car has not been used during your stay the tip of five shillings is quite sufficient. But rules on such points depend on circumstances. If the weather has been bad and the car is in a very muddy state the man will probably have had considerable extra work to bring out your car clean and shining. Remember what you would have had to pay at a public garage and act accordingly.

If you merely pay a call or go to lunch or tea with a friend, and your hostess has a chauffeur who takes the car from you and brings it up to the front door at your departure, a little tip, perhaps two shillings, should suffice.

But such a tip is quite an unnecessary one. The man has done nothing but what he has been paid to do by your hostess. He has done no special or extra work especially for you.

It is always a good thing to keep this in mind whether or no a man whom you are about to tip has performed any direct service for you, extra in any way to what he is paid his wages for, in connection with your car. If he has, a tip is not out of place, if you can afford to give one.

Do not let the idea run away with you that simply because you own and drive a car you must be handing tips to everybody. More than half the tips given are absolutely unnecessary.

There are dozens of cases where people foolishly tip. If your hostess's groom drove you in the dog-cart to the station to catch a train you might think a two-shilling tip all-sufficient. Yet when her chauffeur takes you to the same place in a motor-car you wonder whether he will think five shillings is enough. It is really very absurd. If we have to tip, why not treat the motor-car as we would any vehicle and the chauffeur as we would any groom or coachman ?

There are some people who feel justified, if sent up to town in a friend's car, in giving the chauffeur as a tip the amount of the first-class railway fare for the distance. A tip decidedly should be given, but certainly not so large a one as this, in most cases, would figure out.

If taken to town from a country house, or *vice versa*, and one travels in the car with one's hostess, certainly no tip is necessary ; nor should one be given if one goes for a drive with one's hostess.

Should a friend lend you a car for a day or a drive, a small tip is properly given ; but if a friend lends you a car for a tour of some days, the proper thing is to offer to pay the chauffeur's wages for the week. A tip on the top of this should depend on the manner in which the man serves you.

I have mentioned all these different points because at some time or another they may be actual experiences of the woman who owns and drives her own car.

I do not claim to be an authority on tipping. I distribute a good many gold and silver pieces during a year, but I tip for services rendered and use common sense about the amounts. I get the best of service everywhere.

If every woman who drives her own car followed my rule in this respect the tipping nuisance would not be such a terrible thing after all.

Photo *Elliott & Fry*

MISS ISABEL SAVORY, WHO NOT ONLY DRIVES, BUT REPAIRS HER OWN CARS

DISTINGUISHED WOMEN MOTORISTES

The Englishwoman at the Wheel—Her Skill in Mechanics and Map-reading—The Ladies' Automobile Club—Some Noble Women Motoristes—Successful Competitors—Lady Racers at Brooklands—A "Motor Christening"

THERE is no country in the world—not even France, where the motoring movement received its first real start and its keenest pursuit, nor America, where the fair sex is supposed to receive and to exercise its largest freedom—there is no country in the world in which woman may be seen at the helm of a motor-car so frequently as in England. Whatever the cause—whether it be due to a greater sense of security from annoyance on public roads or simply to superiority of pluck, the fact remains that women in England excel their sisters in other countries as greatly in motoring as in horsemanship.

Almost every woman who can afford it is, of course, a motoriste in the sense that she owns, or has at her disposal, a motor-car. It is not, however, with the ladies whose experience of the pastime is limited to a seat beside or behind the driver that this chapter deals, but rather with those who are accustomed to the task of driving and caring for their cars, and who find a healthful recreation in doing it. Twenty or thirty years ago, two of the essentials to a motorist—some acquaintance with mechanics and the ability to understand local topography—were supposed to be beyond the capacity of a woman's brain. The supposition was simply due to the fact that woman's brain had never had occasion to approach these subjects. Fifty years ago a satirical writer—a man, of course—averred that although instruction in "the use of the globes" was part of the curriculum of every girls' school, no woman could understand, or would try to understand, a road map. If the remark was true when it was written it is

Photo by Keturah Collings

BARONESS CAMPBELL DE LORENTZ, THE FIRST LADY IN BRITAIN TO DRIVE HER OWN CAR

certainly not true to-day. The school-room globes have long been buried in the dust of disuse, but the pastimes of cycling and motoring have made the understanding of maps a necessity to every active gentlewoman; indeed the average woman is probably quicker than the average man in gathering from a map the information which it has to offer.

So with mechanics. If a woman wants to learn how to drive and to understand a motor-car, she can and will learn as quickly as a man. Hundreds of women have done and are doing so, and there is many a one whose keen eyes can detect, and whose deft fingers can remedy, a loose nut or a faulty electrical connection in half the time that the professional chauffeur would spend upon the work.

Incontestable evidence of the practical interest which Englishwomen are taking in motoring is afforded by the existence and prosperity of the Ladies' Automobile Club. This institution was established in 1903. The annual subscription is five guineas, and there is an entrance fee of the same amount. There are nearly four hundred members, most of whom are fully competent to drive their own cars. The club has successfully organised a number of tours in England and on the Continent as well as driving competitions at Ranelagh.

The president of the Ladies' Automobile Club, the Duchess of Sutherland, is the *grande dame* of automobilism in England. The Duchess is an accomplished motoriste, and although in cold weather she prefers to be driven by somebody else, in summer she may often be seen at the wheel. Her latest car is a Mercédès.

Another peeress who drives, and drives well, is the Countess of Kinnoull. The Countess shares her husband's fondness for sport, a great variety of which is provided in the neighbourhood of their beautiful Scottish home at Dupplin Castle, and she finds the motor-car an indispensable adjunct to the full enjoyment of country life.

Lady Wimborne, whose energy and industry are as inexhaustible as those of her brother, the late Lord Randolph Churchill, finds the

THE HONBLE. MRS. ASSHETON HARBORD
Drives a Rolls Royce Car, owns her own balloon "The Valkyrie," and has competea with it in seven races

motor-car an invaluable aid to her useful activities as well as a means of recreation. She has three or four cars, and is an able and confident driver.

Lady Viola Talbot, daughter of the Premier Earl of England, is like her father in the love of sport. Like him she is almost as fond of motors as of horses. She is mistress of the car and its appurtenances, and has driven some thousands of miles at home and abroad.

Among other titled ladies who count their miles by the thousand may be named Lady Beatrice Rawson, a devotee of the small car; Lady Muriel Gore-Brown, the Hon. Mrs. Maurice Gifford, of Boothby Hall, Grantham; Lady Plowden, and the Baroness Campbell de Laurentz. The Baroness has the distinction of being the first lady in Britain to drive and manage her own car. Cars in those days were patterned like high dog-carts and were mostly steamers. The Baroness possesses several photographs of herself and husband, with a groom in the tiger's seat, of the old-fashioned car. To-day these pictures have a very queer look.

A complete list of the ladies who have taken part in motor-car road trials and club competitions would be wearisome to the reader, but a few names may be mentioned. Miss Muriel Hind, one of the few women who favour the motor-bicycle, has won many medals in long-distance trials. Mrs. Herbert Lloyd, who is not only an expert driver, but the inventor of some very ingenious appliances for motor-cars, has done well in open competition with professional male drivers. Miss Daisy Hampson has won prizes with her 120 horse-power Fiat. Mrs. E. Manville has taken part in the Herkomer competition.

No list of distinguished women motoristes would be complete without the names of those who took part in the first race for ladies upon the Brooklands course. The race, which was called the Ladies' Bracelet Handicap, took place in July 1908. There were five starters: Mrs. Locke-King, wife of the founder and owner of Brooklands Racecourse; Miss Muriel Thompson, Miss Christabel Ellis, Miss N. Ridge-Jones, and Mrs. J. Roland Hewitt. Mrs. Locke-King, who started from scratch,

Photo by Arthur Rouselle

MRS. GEORGE THRUPP, ORIGINATOR OF THE MOTOR CHRISTENING

finished a length in front of Miss Muriel Thompson, her speed over the course being at the rate of fifty miles an hour.

There are few lady motoristes who take a keener practical interest in their cars than Miss Isabel Savory. Miss Savory, who has driven many cars, is loud in her praises of her 10 horse-power Cadillac. This car she has always driven and attended to herself, never having employed a chauffeur. She has done all the oiling and adjustments and has put on many a new inner tube by the roadside. She has driven long distances without any companion, and has dealt single-handed and successfully with every emergency that has arisen.

Mrs. George Thrupp, of Cadogan Square, has been driving ever since 1896. Her favourite cars are the Renault and Mors, in which she has toured in Great Britain and on the Continent. She has won prizes in driving competitions. She was, in fact, one of the pioneers of motoring for women, and her youngest boy, Roger, was the first baby to have a " motor christening."

Other names of motoring women that

occur to one are those of Mrs. Edward Kennard, the novelist, who is equally at home at the wheel of a 40 horse-power Napier car and in the saddle of a motor-bicycle; Miss Hunter Baillie, of Cirencester; Mrs. Mark Mayhew, Miss Schiff, Mrs. Claude Paine, Mrs. Nicol, Mrs. Weguelin, Mrs. Charles Jarrott, and Mrs. Edge. No doubt there are other names which at the moment have slipped the memory but which have as good a claim as these to inclusion in the catalogue of distinguished women motoristes. The list is long enough, however, to show the ardour and success with which women have applied themselves to the mechanical details of automobilism.

THE COMING OF THE SMALL CAR

A good Car at a low Price—Lessons of the "Small Car Derby"—Some notable Small Cars and their Cost—Comparatively low Running Expenses—The Car of the People

One of the latest and most notable developments of the motor industry is the prominence of the small car. It is obvious that the number of individuals who can afford to pay from £150 to £230 for a motor-car for purposes either of pleasure or business is enormous in proportion to the number of those who can afford to pay more. Motor manufacturers have never been blind to that fact. It is only in recent years, however, that they have seen their way to meet the demand with satisfaction to their customers and profit to themselves. The small car has come, and its merits are so considerable, the pockets to which it appeals so many, that its popularity is assured. It is not a very rash prophecy to declare that in a few years' time it will be the vehicle most commonly met with on the high road, and its manufacture the mainstay of the motor trade.

In France, where the possibilities—commercial and practical—of the small car were first foreseen, the great motor race of the year, the Grand Prix of the Automobile Club de France, is now preceded by a Grand Prix des Voiturettes, and the result of the 1908 contest is a striking illustration of the speed and reliability of which some of these little vehicles are capable. Of the forty-seven voiturettes which went to the post, twenty completed the course of 286 miles in a little more than seven and a half hours. The winner, a car driven by a single-cylinder De Dion engine and weighing little more than twelve hundredweight, covered the distance in five hours and three-quarters—an average speed of nearly fifty miles an hour; while the second car, a single-cylinder Sizaire, which was only six minutes behind at the finish, covered one of the laps at an average speed of more than fifty-three miles an hour. Speeds such as these are happily not lawful on English roads. I point to them only to illustrate the power that the motor manufacturer has succeeded in obtaining from a single cylinder of less than four inches bore, and the excellence of design and material which has enabled him to produce a little vehicle, weighing a good deal less than a ton, yet capable of withstanding the shocks of rattling over 286 miles of country road at racing speed.

The race for the Grand Prix des Voiturettes and other recent trials have amply demonstrated the speed and reliability of the small car. They have shown that for sums ranging from £150 to £230 the manufacturer can make a car which, for all practical purposes of every-day use upon the road, is the equal in speed and trustworthiness of a car costing from three to five times as much. The 8 horse-power De Dion, which costs £225 15*s*., went through the International Touring-car Trial of 1908 with flying colours. It covered 1800 miles of arduous road work in capital style, and by shedding one of its passengers it even managed to climb the terrible two-mile slope of the Kirkstone Pass. The 9 horse-power Sizaire, the engine already referred to as having accomplished the fastest lap in the Grand Prix, costs 190 guineas. There are many other cars, British and foreign, not less reliable and equally moderate in price: the Phœnix, for example, a twin-cylinder car, costing £140; the Jackson De Dion, costing £220; the Pick, a four-cylinder 14–16 horse-power car, costing only £165; the Rover, costing from 100 to 200 guineas according to engine-power and finish; and the Vauxhall. When it is remembered that cars can often be bought second-hand but in first-class condition for about two-thirds of their original cost, it will be recognised that motoring need not be the exclusive privilege of the very rich.

It is not, however, in the comparison of first cost so much as in that of the cost of maintenance that the small car appeals to the man of moderate means. Generally speaking it may be said that as compared with a full-powered car the small car uses little more than one-third the quantity of petrol per mile travelled, and that its tyres cost only half as much and last twice as long. A gallon of petrol, which will propel a big car 12 or 15 miles, will propel a little Rover or Phœnix from 30 to 40 miles. Here is a statement of the actual cost of running a 6 horse-power Rover for eleven months over nearly 6000 miles of country roads:

	£	s.	d.
Tyre Repairs	2	2	9
Petrol	12	0	0
Oil		18	0
Sundry Repairs	4	12	5
Total	19	13	2

The car belonged to a doctor who had to make frequent stoppages on the way, so that the consumption of petrol was higher than it would have been with continuous travelling. Nevertheless, the cost of running the car works out at about four-fifths of a penny per mile—less than third-class railway fare for one person. This is, no doubt, an exceptionally low figure. Another user of an exactly similar car has found the cost of running 3400 miles to be as follows:

	£	s.	d.
Tyre Repairs	2	13	0
Petrol	6	19	0
Oil and Grease	2	6	0
Sundry Repairs	1	10	0
Charging Accumulators		18	0
Brushes and Waste		10	0
Total .	14	16	0

—almost exactly a penny a mile. To these figures must, of course, be added the cost of licences, insurance, garaging, and an allowance for the depreciation of the car—that is to say, the difference between its first cost and the price at which it could be resold.

In every respect but one the advantages of the small car over its big brother are enormous. Its one drawback is that its accommodation is necessarily smaller. The typical small car is a two-seater; but that is the essence of its economy. Extra seats and extra passengers mean extra weight, and extra weight requires larger engines and more petrol, and entails more wear and tear on tyres, machinery and chassis. It is the weight that runs up the maintenance bill and the prospective purchaser should bear this in mind if he hankers after an extra seat. I may point out, nevertheless, that some of the small cars in the market can be fitted with a detachable rear seat for £6 or £7, and that others have sufficient space for the accommodation of an extra passenger upon the floor. A friend who owns a "two-seater" Sizaire, tells me that it often carries four passengers to the railway station.

It is as a two-seater, however, that the small car bases its claim to popular favour. In the majority of journeys by any sort of private vehicle two is the number for which accommodation is most frequently required. Many owners of large cars have discovered that the occasions on which a two-seater would not serve their motoring purposes are comparatively few. Obviously it is gross extravagance to employ the voracious eater of petrol and rubber upon a service which can be accomplished at a quarter of the cost by a smaller car, at the same speed, with less strain upon the driver and with equal comfort to the passenger. For these reasons the time is at hand when every motor-car owner, however many big cars he may possess, must add to his fleet at least one two-seater for run-about purposes. The large car will be a luxury, the small car will be a necessity—and not only for those who are ordinarily described as wealthy. The time is coming when every man who can afford to go out of town at week ends will find it worth his while to be a motorist, when every suburban house with a rental of over £60 will have its motor shed, and when the small car will be as prevalent upon the country road as the bicycle is to-day.

CAR INDEX-MARKS AND THEIR LOCALE

One sees every day on the streets and roads cars bearing numbers and letters quite unfamiliar. It is advantageous, in many ways, for the motoriste to be fairly well acquainted with some of the more important index-marks. One can easily remember in the United Kingdom that Ireland's index-letters all contain the letter I and Scotland's all the letter S (with two exceptions). England and Wales to date, with very few exceptions, use up the letters A, B, C, D, E and F. London has now four distinct index-marks and no doubt will add to them as the increase in cars may demand.

All motor-cars must have an index-mark and a registration number, excepting those owned personally by his Majesty the King.

White letters and figures on a black plate are for privately owned cars. Trade vehicles use coloured figures and letters, usually red on a white plate. Trade vehicles usually also have additional letters which are granted them as a trade-mark or for trade purposes. But the index-mark or letter of their locale must, in all instances, be placed first on the plate.

Registration need not be effected in the same district in which the car is owned, so that, with some trouble, an owner can have practically any letter she likes on her car.

The fee for registration, £1, need be paid only once on any one car, excepting on change of ownership, when a fee of 5*s.* is payable. But with this change of ownership the index-mark and registration number remain the same. If a new index-mark and registration number are wanted, the existing ones can be cancelled and the car re-registered, in any locality, on payment of the full fee.

The following are the index-marks in use in the United Kingdom

ENGLAND AND WALES

Index-Mark.	Locale.
A	London (also LB, LC and LN)
AA	Southampton County Council
AB	Worcestershire
AC	Warwickshire
AD	Gloucestershire
AE	Bristol
AF	Cornwall
AH	Norfolk
AJ	Yorkshire, N. Riding
AK	Bradford
AL	Nottinghamshire
AM	Wiltshire
AN	West Ham
AO	Cumberland
AP	Sussex, East
AR	Hertfordshire
AT	Kingston-on-Hull
AU	Nottingham
AW	Shropshire
AX	Monmouthshire
AY	Leicestershire
B	Lancashire
BA	Salford
BB	Newcastle-on-Tyne
BC	Leicester
BD	Northamptonshire
BE	Lindsey
BF	Dorsetshire (also FX)
BH	Buckinghamshire
BJ	Suffolk, East
BK	Portsmouth
BL	Berkshire
BM	Bedfordshire

Index-Mark.	Locale.
BN	Bolton
BO	Cardiff
BP	Sussex, West
BR	Sunderland
BT	Yorkshire, East Riding
BU	Oldham
BW	Oxfordshire
BX	Carmarthenshire
BY	Croydon
C	Yorkshire, West Riding
CA	Denbighshire
CB	Blackburn
CC	Carnarvonshire
CD	Brighton
CE	Cambridgeshire
CF	Suffolk, West
CH	Derby
CJ	Herefordshire
CK	Preston
CL	Norwich
CM	Birkenhead
CN	Gateshead
CO	Plymouth
CP	Halifax
CR	Southampton Borough
CT	Kesteven
CU	South Shields
CW	Burnley
CX	Huddersfield
CY	Swansea
D	Kent
DA	Wolverhampton
DB	Stockport
DC	Middlesbrough
DE	Pembrokeshire

INDEX-MARK.	LOCALE.
DF	Northampton Borough (also NH)
DH	Walsall
DJ	St. Helens
DK	Rochdale
DL	Isle of Wight
DM	Flintshire
DN	York
DO	Holland, Lincs.
DP	Reading
DR	Devonport
DU	Coventry
DW	Newport, Mon.
DX	Ipswich
DY	Hastings
E	Staffordshire
EA	West Bromwich
EB	Isle of Ely
EC	Westmorland
ED	Warrington
EE	Grimsby
EF	West Hartlepool
EH	Hanley
EJ	Cardiganshire
EK	Wigan
EL	Bournemouth
EM	Bootle
EN	Bury
EO	Barrow-in-Furness
EP	Montgomeryshire
ET	Rotherham
EU	Breckonshire
EW	Huntingdonshire
EX	Great Yarmouth
EY	Anglesey
F	Essex
FA	Burton-on-Trent

INDEX-MARK.	LOCALE.
FB	Bath
FC	Oxford
FD	Dudley
FE	Lincoln
FF	Merionethshire
FH	Gloucester
FJ	Exeter
FK	Worcester
FL	Peterborough
FM	Chester
FN	Canterbury
FO	Radnorshire
FP	Rutlandshire
FR	Blackpool
FT	Tynemouth
FX	Dorsetshire (also BF)
FY	Southport
H	Middlesex
HA	Smethwick
J	Durham, County
K	Liverpool
L	Glamorganshire
LB	London (also A)
LC	,,
LN	,,
M	Cheshire
MN	Isle of Man
N	Manchester
NH	Northampton (also DF)
O	Birmingham
P	Surrey
R	Derbyshire
T	Devonshire
U	Leeds
W	Sheffield
X	Northumberland
Y	Somersetshire

SCOTLAND

All cars bearing on their index-marks the letter S can, at a glance, be put down as Scottish, for Scotland alone has a right to the use of this letter:

Index-Mark.	Locale.
AS	Nairnshire
BS	Orkney
DS	Peeblesshire
ES	Perthshire
G	Glasgow
HS	Renfrewshire
JS	Ross and Cromarty
KS	Roxburghshire
LS	Selkirkshire
MS	Stirlingshire
NS	Sutherlandshire
OS	Wigtownshire
PS	Shetland
RS	Aberdeen City
S	Edinburgh
SA	Aberdeen County
SB	Argyllshire
SD	Ayrshire
SE	Banffshire
SH	Berwickshire
SJ	Buteshire
SK	Caithness-shire
SL	Clackmannanshire
SM	Dumfriesshire
SN	Dumbartonshire
SO	Elginshire
SP	Fifeshire
SR	Forfarshire
SS	Haddingtonshire
ST	Inverness-shire
SU	Kincardineshire
SV	Kinross-shire
SW	Kircudbrightshire
SX	Linlithgowshire
SY	Midlothian
TS	Dundee
US	Govan
V	Lanarkshire
VS	Greenock
WS	Leith
XS	Paisley
YS	Partick

IRELAND

The inclusion of the letter I on a car's index-mark stamps that car at once as Irish, for the use of this letter is confined to Ireland:

Index-Mark.	Locale.
AI	Co. Meath
BI	Co. Monaghan
CI	Queen's County
DI	Co. Roscommon
EI	Co. Sligo
FI	Tipperary, North
HI	Tipperary, South
IA	Co. Antrim
IB	Co. Armagh
IC	Co. Carlow
ID	Co. Cavan
IE	Co. Clare
IF	Co. Cork
IH	Co. Donegal
IJ	Co. Down
IK	Co. Dublin
IL	Co. Fermanagh
IM	Co. Galway
IN	Co. Kerry
IO	Co. Kildare
IP	Co. Kilkenny
IR	King's County
IT	Co. Leitrim
IU	Co. Limerick
IW	Co. Londonderry
IX	Co. Longford
IY	Co. Louth
IZ	Co. Mayo
JI	Co. Tyrone
KI	Co. Waterford
LI	Co. West Meath
MI	Co. Wexford
NI	Co. Wicklow
OI	Belfast
PI	Cork
RI	Dublin
TI	Limerick
UI	Londonderry
WI	Waterford

FRANCE

In France the index-numbers are divided among sixteen sections, including Algeria, which are called *Arrondissements minéralogiques.* Some of these sections contain as many as eight departments or counties. The majority have single letters. Paris has five sets of single letters. It is necessary for a motoriste from the United Kingdom, when taking her car into France, to affix a French index-mark above her British one. This mark and number will be given her at the point of debarkation on registering the car and on passing a pleasant and easy little examination in the art of driving. This test consists in driving round a square or up and down the street for about a quarter or half a mile, turning a few corners *en route.* The French index-marks are as follows :

Index-Mark.	Locale.
A	*Alais* (Gard, Hérault, Lozère, Ardèche)
A-L	Algérie
B	*Bordeaux* (Départements du sud de la Garonne, Charente - Inférieure, Dordogne, Lot - et - Garonne)
C	*Châlon-sur-Saône* (Côte-d'Or, Jura, Ain, Doubs, Yonne, Saône-et-Loire)

Index-Mark.	Locale.	Index-Mark.	Locale.
D	*Douai* (Nord, Aisne)	N-O	*Nancy* (Départements de l'Est, including the Aube)
E	Paris (also G, I, U and X)	P-K	*Poitiers* (Départements du sud de la Loire, including the Loiret)
F	*Clermont-Ferrand* (Puy-de-Dôme, Haute-Loire, Cantal, Allier, Nièvre)	S	*St.-Etienne* (Rhône, Loire)
G	Paris (also E, I, U and X)	T	*Toulouse* (Languedoc, including Tarn and Lot)
H	*Chambéry* (Savoie, Haute-Savoie, Isère, Drôme, Basses-Alpes)	U	Paris (also E, G and I)
I	Paris (also E, G, U and X)	X	Paris
L	*Le Mans* (Sarthe, Départements de la Bretagne, Mayenne)	Y or X	*Rouen* (Seine-et-Oise, Seine-et-Marne, Eure, Eure-et-Loire, Seine-Inférieure, Orne, Calvados, Manche)
M-V	*Marseille* (Départements de la Côte, Corse)		

GERMANY

The motoriste from the United Kingdom can take her car into Germany and drive it there without having to put on a new number. Germany willingly accepts the British index-mark already on the car. Index-marks in Germany are allotted in twenty-six different sections, one of which, Prussia, is sub-divided into twelve provinces and the city of Berlin.

In nearly all the distinct kingdoms and duchies an attempt has been made to utilise the initial letter of that section, thereby making recognition of the locale of the car quicker.

In Prussia the mark is the number I in a Roman figure followed by letters of the alphabet. In many other provinces numbers in Roman figures are also used, the figure in most instances being followed by a letter of the alphabet.

The following are the German police index-marks for cars (*Kraftfahrzeuge*):

Index-Mark.	Locale.
	Prussia
I A	Berlin
I C	East Prussia
I D	West Prussia
I E	Brandenburg
I H	Pomerania
I J	Posen
I K	Silesia
I M	Saxony
I P	Schleswig-Holstein
I S	Hanover
I T	Hesse-Nassau
I X	Westphalia
I Z	Rhine Province

Index-Mark.	Locale.
Figures II and letters A, B, C and so on	Bavaria
Figure only, I II, III, IV and V	Saxony (Kingdom)
Figures III and letters A, B, C and so on	Würtemburg

Car Index-Marks

Index-Mark.	Locale.
Figure IV and letters	Baden
Figure V and letters	Hesse
M I	Mecklenburg-Schwerin
S	Saxony (Grand Duchy)
M II	Mecklenburg-Strelitz
O	Oldenburg
B	Brunswick
SM	Saxe-Meiningen
SA	Saxe-Altenburg
KG	Saxe-Coburg-Gotha
A	Anhalt
SR	Schwarzburg - Rudolstadt
SS	Schwarzburg-Sondershausen
W	Waldeck
RA	Reuss (old line)
RJ	Reuss (young line)
SL	Schaumburg-Lippe
L	Lippe
HL	Lübeck
HB	Bremen
HH	Hamburg
Figure VI and letter A, B and so on	Alsace-Lorraine (Elsass-Lothringen)

THE MOTOR WOMAN'S DICTIONARY

ACCELERATOR.—A device, operated by a pedal, for increasing the speed of the engine, either by suspending the controlling action of the governor or opening the throttle.

ACCUMULATOR.—An apparatus for storing electricity. *See* Battery.

ACETYLÈNE.—An inflammable gas giving a brilliant light. It is commonly produced by adding water to carbide of calcium.

AMMETER.—An instrument used for measuring the number of ampères in an electrical circuit. *See* Ampère.

AMPÈRE.—The unit of measure of the quantity of current flowing through an electrical circuit. *See* Volt.

AXLES.—The horizontal shafts or girders by which the weight of carriage is transferred to the road wheels and at the ends of which the wheels revolve. A "live" axle is one which revolves with, and by which the power of the engine is communicated to, the driving-wheels.

BACKFIRE.—A premature explosion of the gaseous mixture in the cylinder. When it occurs while the starting-handle is being turned its effects are distinctly unpleasant to the operator.

BATTERY.—An arrangement of two or more cells either for the production or the storage of electricity. In the former case it is known as a primary battery; in the latter case it is known as a secondary battery, a storage battery, or an accumulator. *See* Cell.

CELL, ELECTRICAL.—A chemical apparatus for the production or storage of electricity. Two or more cells electrically connected constitute a battery.

CIRCULATING PUMP.—The pump which forces the water through the radiator to ensure rapid cooling.

BEARINGS.—The cradles or surfaces upon which the moving parts of machinery are supported. They always require lubrication.

BELT.—A flexible band sometimes used instead of gearing to transmit the power of the engine to the driving-wheels.

BEVEL GEAR.—A gear consisting of cog-wheels with bevelled or sloping surfaces.

BIG END.—The end of the connecting-rod which grasps the crank. *See* Connecting-rod.

BRAKE.—A device for stopping or checking the motion of the car by the application of friction to one or other of the moving parts. A motor-car should have at least three good brakes applicable by the hand and foot of the driver.

BRAKE HORSE-POWER. *See* Horse-power.

CAM.—A revolving disc of irregular shape designed to impart a variable motion to some other piece of mechanism—such as the spindle of a valve—in contact with it.

CARBIDE OF CALCIUM.—A compound of chalk and coke which, when brought into contact with water, gives off the acetylene gas used for lighting.

CARBURETTOR.—The apparatus for regulating the rate of evaporation of the petrol and the proportions of its mixture with air. It consists generally of a float chamber and a mixing-chamber.

CHAIN.—In motor-cars chains are sometimes used, as in the bicycle, for the transmission of power from one part of the mechanism to another.

CHANGE-SPEED GEAR.—The arrangement of shafts and toothed wheels by which the speed of the road wheels is altered without alterations in the speed of the engine.

CHASSIS.—The entire framework and mechanism of the car—engines, wheels, axles, &c.—without the body or seating accommodation.

CLUTCH.—A device for connecting the motive machinery with the driving-wheels at the will of the driver.

COIL.—*See* Induction Coil.

COMMUTATOR.—An appliance for enabling the driver to advance or retard the spark which ignites the mixture in the cylinder.

COMPRESSION.—This word in motoring invariably applies to the compression of the gaseous mixture in the cylinder. The efficiency and economy of the motor depend greatly upon the degree of compression at the moment of ignition.

CONNECTING-ROD.—The rod which connects the piston with the crank of the engine.

CUT-OUT.—A device for diverting the exhaust gases directly into the air instead of compelling them to pass through the silencer. The "cut-out" is seldom used except in racing. It accelerates the engine at the cost of an appalling noise.

CYCLE.—*See* Otto Cycle.

CYLINDER.—The cast-iron chamber in which the petrol gas is compressed and exploded and in which the piston travels to and fro.

DENSIMETER.—An instrument for testing the specific gravity (*i.e.*, the weight as compared with water) of petrol.

DIFFERENTIAL GEAR.—The gear by which power is communicated to the driving-wheels in such a way that it is taken up automatically by either or both of them according to their respective requirements at the moment.

DRY CELL, OR BATTERY.—A cell, or battery of cells, which works without liquid. Dry cells are generally used for motoring purposes in preference to cells containing solution.

DYNAMOMETER.—An instrument for testing the amount of power expended by mechanical or animal effort. The dynamometer used for testing motors is often called the "pony brake." *See* Horse-power.

ELECTRIC IGNITION.—The ignition of the explosive mixture in the cylinder is effected by an electric spark caused by forcing an electric current through the sparking-plug. The current is obtained (1) from an accumulator or a primary battery, in which case its pressure is raised to the required voltage by passing it through an induction coil; or (2) from a magneto-electrical instrument—which is very much like a dynamo on a small scale, and is driven by the motor.

EXHAUST.—The gaseous products of combustion expelled from the cylinder during the scavenging stroke of the piston.

FAN.—A rotary fan, driven by the motor, is often employed to increase the current of air passing through the radiator and thus to assist in cooling the water.

FEED.—The method of conveying the petrol from the main tank to the carburettor. If the tank is higher than the carburettor, the petrol will pass by gravity. Otherwise it must be lifted by pressure. The exhaust is generally employed for this purpose, a hand-pump being fitted to furnish the necessary pressure for starting.

FLY-WHEEL.—As only one of the four strokes of the motor is a working stroke, a heavy fly-wheel is necessary to carry the piston through its cycle and promote easy running.

FRAME.—The structure which carries the machinery of the car.

GARAGE.—A stable for motor-cars.

GEARS.—*See* Bevel Gear, Change-speed Gear, Differential Gear.

GOVERNOR.—The appliance which automatically regulates the speed of the engine, usually by checking the volume of mixture admitted to the cylinder.

GRADIENT.—The inclination or slope of a road.

GRAVITY FEED.—*See* Feed.

HORSE-POWER.—Boulton and Watt calculated that a London dray-horse was capable of work equivalent to lifting 33,000 lb. one foot high in one minute, and this task—technically described as 33,000 foot-lb. per minute—has been accepted as the "unit of horse-power" for the measurement of mechanical work. The power of a petrol motor depends upon many factors—diameter of cylinder, speed of working, quality of mixture, compression, cooling-surfaces, &c.—some of which vary from moment to moment in practice. The only real means of measuring it is by the dynamometer or pony-brake, which records the power actually available for useful work. Horse-power so measured is called brake horse-power. For purposes of competitions the Royal Automobile Club use the following formula for rating the comparative power of petrol motors. Multiply the diameter of the cylinder in inches by itself and again by the number of cylinders. The product divided by 2½ is the nominal horse-power.

IGNITION.—*See* Electric Ignition.

INDUCTION-COIL.—An apparatus for intensifying the pressure of the electric current. Used in motor practice as a part of the system of high-tension ignition.

INSPECTION PIT.—A pit or well, generally situated in or near the garage, to facilitate the examination and repair of the chassis of a car.

LICENCES.—Licences must be taken out (1) by the owner of a car. This licence costs from two to five guineas according to the weight of the car; and (2) the driver of a car. This costs five shillings only.

LIMOUSINE.—A large covered car.

LIVE AXLE.—*See* Axles.

LUBRICATION.—The application of oil, grease or other substances suitable for the reduction of friction between sliding surfaces.

MIXTURE.—The explosive charge of petrol and air admitted to the cylinder. *See* Carburettor.

OTTO CYCLE.—Nearly all petrol motor-car engines work upon the system invented by Otto in 1876 and known as the Otto Cycle.

The cycle consists of the successive operation of induction, compression, explosion and scavenging, there being thus only one working stroke in four—that is in every two revolutions of the fly-wheel.

Over-Heating.—An accident generally resulting from deficient water circulation or insufficient lubrication. Unless speedily remedied it may result in serious damage to the motor. *See* Seizing.

Panne.—A French word, meaning "a breakdown."

Petrol.—A highly inflammable and volatile spirit distilled from petroleum. Seven pints of petrol weigh about as much as ten pints of water. Its vapour is heavier than air.

Piston.—The disc which slides up and down in the cylinder, and communicates the force of the explosion to the connecting-rod and crank.

Piston Rings.—Cast-iron split rings, fitted in grooves round the piston to make a gas-tight joint between the piston and the walls of the cylinder.

Pony-brake.—*See* Dynamometer.

PRESSURE FEED.—*See* Feed.

RADIATOR.—A device for cooling the water circulating round the cylinders by exposing it to a large surface in contact with free air.

SIDE-SLIP.—An accident liable to occur on greasy roads, but of less frequent occurrence since the introduction of "non-skidding" devices.

SILENCER.—A box or chamber designed to reduce the noise caused by the escape of the exhaust gases.

SPARES.—Duplicate parts of machinery carried in case of a breakdown.

SPARKING-PLUG.—A porcelain plug, carrying the electric wires, which is screwed into the combustion chamber of the cylinder. At the end of the plug within the cylinder are platinum points in connection with the wires. The current jumping from point to point makes the spark which fires the charge.

THROTTLE.—The control of the volume of mixture supplied to the engine.

TIMING-GEAR.—The gear which controls the times at which the valves of the engine open and close, and at which the charge is fired.

TORQUE.—The twisting effort of rotation.

UNIVERSAL JOINT.—A flexible joint which permits the transmission of power from one shaft to another in any direction.

VOLT.—The unit of measure of electrical pressure. It is the measure of the quality of the flow as compared with ampère which is the measure of quantity.

WEBLEY & SCOTT Ltd.

A necessity for every motorist travelling at night and in out-of-the-way places, the Webley Automatic Pocket Pistol.

The chief features of these weapons are their simplicity of construction, strength of mechanism, lightness and portability. In these three features they are superior to any other Automatic Pistols on the Market.

WEAMAN ST., BIRMINGHAM, and 78, SHAFTESBURY AVENUE, LONDON, W.

GENERAL ELECTION 1910.

All Parliamentary Candidates
insist on having

"SHELL" MOTOR SPIRIT

WHY ?

Because it is the
SUREST way of getting there,
Because it is the
QUICKEST way of getting there,
Because it is the
SAFEST way of getting there.

"The Light Touch of a Gentle Lady"

will drive an "Argyll" car. Everything is simplicity, everythin
"just there;" the driver has only to look ahead.

The "Argyll" cars will "go-anywhere-and-do-anything," and t
excel in reliability and beauty of design.

Send for New Art Catalogue "B," which tells intending purchasers all they require to kno

London Agents: "ARGYLLS, LONDON," Ltd., 17, Newman St., Oxford St., W.

ARGYLL MOTORS, LTD., Argyll Works, Bridgeton, GLASGOW.